WHAT CAN THIS AMAZING BOOK DO FOR YOU?

A VITAL HEALTH MESSAGE TO THE NATION YEARS AHEAD OF OUR TIME

A general dedication to all mankind who desire to live far beyond their normal years.

A miraculous transformation could occur within you, when you discover that which is perfect for man.

ARE YOU CONCERNED ABOUT:

Arthritis
Hardening of the Arteries
Kidney Stones
Gall Stones
Cataracts
Glaucoma
Loss of Hearing
Diabetes
Obesity
Emphysema

THE CHOICE IS CLEAR

by Dr. Allen E. Banik

24th Printing (1,370,000) copies sold. ISBN 0-911311-31-9

ACRES U.S.A.
P.O. Box 91299 • Austin, Texas 78709
(512) 892-4400 • fax (512) 892-4448
info@acresusa.com • www.acresusa.com

INTRODUCTION

Dr. Allen E. Banik's insatiable quest for knowledge, and intense desire to trace all chronic and fatal diseases to a common cause, somehow caught the attention of Art Linkletter, who wanted some health scientist to visit the long-lived people of Hunza, and probably bring back the secret of their amazing spans of life. Dr. Banik startled the world with his report to the nation.

As Mr. Linkletter stated in the book, *Hunza Land*, (by Dr. Banik), "We could easily have asked for someone at random in our studio audience—but for an examination of the strange people in the far-away Himalayas, we needed a man with a scientific background and a deep interest in the study of geriatrics.

"Dr. Banik proved to be that man, an optometrist living in Kearney, Nebraska, whose name came up repeatedly in our quest for a hobbyist who pursued the provocative study of old age.

"We flew a *People Are Funny* scout to his home town to interview him to make sure he was the right man for this kind of stunt. The report came back, 'great.' Dr. Banik was the ideal guest: enthusiastic, cooperative, curious, happy, and an intelligent observer.

"He and Hunza made one of the most successful episodes in our show's history. My only regret was that time-on-the-air did not permit the full story of his eventful trip to be told. Happily, his book, *Hunza Land*, fills in the fascinating details and unfolds the complete story of one of the most unusual postage-stamp countries in the world."

In his book, *The Choice Is Clear*, Dr. Banik continues his findings in *Hunza Land*, only now to shock the world, with the one finding he nearly missed—one of the real secrets of longevity.

THE CHOICE IS CLEAR

The amazing discovery I am about to unfold to you had its beginning some 40 years ago. Just why I did not receive the impact then I will never know. A few short months ago the truth finally revealed itself to me. At first I found it difficult to comprehend, but little by little the facts began to blend until I saw it all in its entirety, and *The Choice Was Clear*.

Only One Possible Cause For All Aging Diseases?

Imagine one common possible cause for nearly all diseases! Can it be that arthritis, kidney stones, gall stones, arteriosclerosis (hardening of the arteries), enlarged hearts, emphysema, obesity, constipation, cataracts, glaucoma, diabetes, all stem from one common cause? Unbelievable! But possibly very true!

Can This Be True?

We all use it every day—from the baby's first cry to the old man's last breath. Water! The one element you need the most and suspect the least. If this shocks you, do not be surprised, as millions of Americans are still unaware of this "aging process," and indulge in it freely. Multitudes suffer from all the above diseases, and unknowingly contribute to this cause daily. Somehow the shots, pills, vitamin capsules and balanced diets add only brief spurts to well-being.

The arthritis can get a little worse, the arteries a little harder, the gall stones and kidney stones a little larger, the eyes a little dimmer, the hearing more complex, sugar starts to appear in the urine, digestion slows down, constipation calls for stronger laxatives—and waist lines get a little larger. It's a never-ending cycle of feeling just a little worse than the day before.

There Is Hope Ahead

Let me cheer you up—there is good news in store for you—happy years are now ahead of you—far into your nineties—even more. Your stiff joints may begin to loosen, arteries become more elastic, blood pressures can become more normal, heart action firmer and more powerful. Your lungs can expand with fresh air—kidney and gall stones can gradually dissolve, never to appear again. That lost appetite can also reappear, and those aggravating stomach pains somehow can vanish.

Dr. Brown Landone And Captain Diamond Proved It

Let me quote a few authenticated case histories—just to prove what has already taken place in the lives of two famous individuals. One was an associate named Dr. Brown Landone who lived to the splendid age of 98. His patient, a Captain Diamond, lived vigorously to the fantastic age of 120 years.

Dr. Landone had a heart condition at the age of 17. Today it would have meant a heart transplant or other dangerous surgery. His doctor allotted him three months to live at the most. His heart chambers and valves were coated with mineral deposits—much like his mother's teakettle. Dr. Landone was determined to live. After devising a special treatment (I will go into details later), he returned for his second heart examination two years later. His doctor pronounced his heart *perfect*. All aglow with life, young Landone enrolled in a medical college and received the degree of Doctor of Neurology. He practiced this profession for 50 years, then turned his talents to writing. His average working day was from 18 to 20 hours. At the age of 98, he could have been mistaken for a man of 50!

The Amazing Age of Captain Diamond

During Dr. Landone's years in practice, he met Captain Diamond, who was bed-ridden with arthritis and hardening of his arteries at the age of 70.

Dr. Landone suggested his treatment. Five years later he was teaching classes in physical culture. At the age of 103 he walked from Sacramento to New York. At the age of 110 he danced all night with a 16-year-old. He died at the fantastic age of 120!

The Secret of Longevity

"Water" is what I will be writing about, from the very first page to the last word. I will attempt to show facts about water that will startle you. And from now on, until the day you draw your last breath, you will remember what I have written on these pages.

All Water Is Not The Same

There are at least nine different kinds of water. Some kinds can harden your arteries, form gall stones and kidney stones, bring on early senility, and all the diseases I mentioned earlier. Other kinds of water works in reverse. What one type of water carries into the system, the other carries out. Let me classify these nine kinds of water. They are *hard water, raw water, boiled water, soft water, rain water, snow water, filtered water, de-ionized water* and *distilled water*. All are kinds of water—but remember this: only one of these nine kinds of water is good for you. Let me describe them briefly.

Hard water is water containing excessive lime salts, that is, carbonates and sulfates of calcium and magnesium. Sodium, iron, copper, silicon, nitrates, chlorides, viruses, bacteria, chemicals, and many other harmful inorganic minerals and chemicals may also be present. Most all the water we drink comes from public water systems or private wells. These are hard waters. Any water that has run through, or over, ground is hard water to some degree. The longer it filters through the soil, the harder it gets and the more harmful it may become for you.

Some cities take water from rivers and lakes or mountain reservoirs. They call it soft water. But soft water from such sources is soft only in comparison to water which is harder.

I recall that each year the earth's rivers carry to the sea five billion tons of dissolved minerals and other unnumbered millions of tons of carbon compounds and factory pollutants. *"And ocean life is on the decrease,"* according to Jacques Cousteau.

A False Type Of Water

There are also the hard waters which are called "mineral waters"—that is, waters from certain mineral springs—well known for their medicinal effects. Practically all kinds of "bottled" water is *hard water*.

The reason mineral waters have this so-called medicinal effect is because the body tries to throw off the excess minerals which invade it as intruding foreign deposits. Actually, a form of dysentery exists, as the organism tries to eliminate and expel this unusual condition. To subsist on this type of water could be detrimental.

Raw water is water which has not been treated in any way. It may be hard or soft—as hard as lime water, or as soft as rain water. Raw water contains millions of viruses and bacteria, and is densely inhabited in every drop. Chemicals dumped into our rivers may cause cancer, according to the Environmental Protection Agency.

True, properly chlorinated water kills most germs and viruses, but it can also kill cells in our bodies. The only safe rule to follow individually is to distill drinking water, which will eliminate both chlorine and pathogenic organisms.

Boiled Water Is Not The Answer

Boiled water is advised by some health authorities. Boiling removes none of the *inorganic* minerals, although it does kill the bacteria in raw water if boiled at least 20 minutes.

But the dead bodies of these germs are carried into the system when the boiled water is used. Such dead materials furnish a fertilized soil for rapid and lusty propagation of germs already in the body. By drinking boiled water one may avoid live disease germs, but still takes on bacterial soil for the growth of other bacteria.

While *raw water* is an aquarium filled with deadly micro-organisms, *boiled water* is a graveyard of dead germs.

The Impurity Of Rain Water

Rain water, which has been distilled by the heat of the sun, should contain no mineral matter and no germs. But when it falls from the clouds as rain, it falls through air filled with bacteria, dust, smoke, chemicals, mud and minerals. By the time it reaches the earth as rain water it is so saturated with decaying matter, dirt and chemicals that its color becomes a yellowish-white.

If rain water is allowed to stand it becomes filthy because of the rotting animal matter in it. I discuss it only because some "nature faddists" are so ignorant that they advise the use of rain water on the basis that it is "natural."

Snow Is Unclean

Snow water is melted snow. It, too, picks up minerals, chemicals and radioactive fallout such as Strontium 90. Snow is frozen rain. Freezing does not destroy bacteria. Snow looks white and clean, but it contains as many germs, minerals and pollutants as rain water. Remember the dirt left from melting snow in the spring? How the dirt of what has melted forms on top of the snow which is not yet melted! Try melting the cleanest snow, then note the filth in the water. Each snowflake contains some form of air pollution.

Filtered Water Can Be Dangerous

Filtered water is water which has passed through a very fine strainer, activated carbon or some other mechanical barrier. The use of filtered water is still rather popular. Some people think that water which has passed through a filter is purified. They believe that the filter keeps the waste substances and disease germs of the water in the filter. While it is true that chlorine, some suspended substances and many synthetic chemicals are removed in the filter by filtering, there is no filter made which can prevent bacteria or viruses from passing through its fine meshes.

Each pore of the finest filter is large enough for millions of germs to pass through, even those containing silver!

Moreover, decaying matter collects on the bottom of every filter. This forms an excellent breeding ground for bacteria. After a filter has been used for a few days, the filtered water often contains more disease germs than the water which is put into the filter. Bacteria are multiplied by the millions by the collected wastes at the base of some filters, and washed through with filtered water, according to Dr. Brown Landone.

De-ionized Water Can Cause Virus Infections

Water processed by the de-ionized method effectively removes minerals, and compares to distilled water in this respect. However, it does become a breeding ground for bacteria, pyrogenic matter and viruses. The fault in this system lies in the resin beds which can become notorious breeding grounds. Therefore it is not wise to have this possibility exist in your drinking water.

Furthermore, deionization does not remove synthetic chemicals such as herbicides, pesticides, insecticides or industrial solvents.

The Miracle Of All Water—Distilled Water!

Distilled water is water which has been turned into vapor, so that virtually all its impurities are left behind. Then, by condensing, it is turned back to "pure" water. Distillation is the single most effective method of water purification. It is God's water for the human race.

In a manner of speaking, distillation is nature's way. The weather of the world is created in the tropics, where half the heat reaching the earth falls on land and water masses. Here heat energy is stored within water vapor through the process of evaporation, nature's distillery. When the jet streams return ocean water to inland areas, they do so without sea salts and minerals, all of which have been left behind.

Nowadays you will hear a lot about reverse osmosis. In this process, water is purified by forcing a portion of the raw water through a semi-permeable membrane. Reverse osmosis removes a high percentage of the dissolved solids as well as other contaminants, and when new the result often approaches the purity of distilled water. The degree of purity in any case varies widely, depending on the types and conditions of the equipment used, much as with filter equipment, and the effectiveness lessens with use. Sometimes drastically!

Nature's Natural Distilling Plant

Distilling water turns it into vapor, and then through condensation, back again into pure water. Rising vapor cannot carry minerals and other dissolved solids—it will not carry disease germs, dead or alive. The secret is that the vapor rises between all the suspended particles and chemicals in the air. When this condensation occurs as falling rain, it picks up airborne pollutants! Not so in a vented distiller where most of them are eliminated.

If pure distilled water is boiled in a teakettle, no calcium or minerals of any kind will collect to coat the inside of the kettle, even though you used the same kettle for ten years.

The World's Greatest Secret

Distilled water then is water of the purest kind. It is odorless, colorless and tasteless.

The divine purpose of water is to regulate temperature and to act as a solvent. In nature, water in evaporation is so fine that your eye cannot perceive it as it is drawn up into the clouds. Then it falls as rain, keeping the earth from being parched and burned.

As a solvent, it dissolves rocks and soil. It figures in the transport of nutrients into plant life.

In the human body, water fills similar functions. It regulates the temperature of the body by helping take off extra heat resulting from an intake of some 3,000 calories of food each day. Water keeps the body from burning up. It carries waste products from the body.

The Greatest Function Of Distilled Water

Distilled water acts as a solvent in the body. It dissolves food substances so they can be assimilated and taken into every cell. It dissolves inorganic mineral substances lodged in tissues of the body so that such substances can be eliminated in the process of purifying the body.

Distilled water is the greatest solvent on earth—the only one that can be taken into the body without damage to the tissues.

By its continued use, it is possible to dissolve inorganic minerals, acid crystals, and all the other waste products of the body without injuring tissues.

Inorganic Mineral Deposits Must Be Removed

For purification, distilled water is the solvent of choice. After mineral deposits are dissolved, gentle muscular exercise can force the dissolved poisons and wastes from the tissues into the blood so that blood can carry wastes to the excretory organs. Then, these organs can squeeze and pour the wastes out of the body.

Remember that great scientists now not only admit but assert that all old age, and even death—unless by accident—is due to waste poisons not washed out of the body. The legendary Dr. Alexis Carrel made heart tissue apparently immortal by regularly washing away the wastes of the cells.

Now you have learned about the nine different types of water. Let me repeat them again so you will never forget them: *hard water, soft water, raw water, boiled water, rain water, snow water, filtered water, de-ionized water* and *distilled water*. It is wise to remember these nine different types of water, but there is only one water you must always remember—*distilled water*! Dr. Paul Bragg, a life-time advocate of distilled water, lived to age 94, the span of life others only lecture about. He and others with track records say this is your lifeline to a long, happy, carefree span of life.

Your Secret Enemy At Work

Hard water then is your greatest enemy. It will destroy every fond hope you have by striking you down when you should be enjoying the fruits of your labors. Crippled joints, repeated surgeries, enlarged hearts, hardened arteries, gall stones, kidney stones, hearing problems, forgetful minds, all draw your activities from the great out-of-doors into creaking rocking chairs, and finally into bed-ridden old people's homes. Why all this, when the body and mind was divinely designed to live forever! Dr. Alexis Carrel proved that! Captain Diamond lived a long merry life.

Let us trace the course of hard water, the cause of our ills. It begins in the clouds. As it falls it collects minerals and poisonous chemicals. It is good water, except for what it picks up as it falls. When it reaches the ground it is divinely designed to do one thing, to pick up minerals. These inorganic minerals are wonderful, but only for plants. As the water picks up minerals, it distributes them evenly to plant life. Plants would never grow if these minerals were not supplied to their roots. Water then acts as a carrier of some minerals, namely anion nutrients. But alas, chemicals have now polluted our pure air. To clean the air, and make it fit to breathe, nature designed water to absorb these poisons.

The Fault Lies In Inorganic Minerals

We have never realized that these collected minerals from the air and in the ground are all *inorganic* minerals, which cannot be assimilated by the body unless they are chelated into organic molecules. The *only* minerals the body can utilize are the *organic* minerals. All the others are foreign minerals and must be disposed of or eliminated. Herein lies the problem we have never suspected as the possible *cause of nearly all our aging diseases*.

Only Plants Need Inorganic Minerals

Water, as a carrier of minerals, is fine for the plant. The plant in turn converts the *inorganic* to *organic* minerals. Now, our bodies can assimilate them. We have worked this process in reverse. We consume hard water saturated with calcium, magnesium, iron, copper, silicon, not realizing the body is unable to assimilate these nutrients efficiently. So nature tucks them in the joints as arthritis, in the intestinal walls as constipation, and along the arteries, causing them to harden. The kidneys and liver roll up the mineral deposits into little stones until they get too large for the ducts. Sometimes the "filters" of the kidneys become so mineral clogged that kidney transplants become necessary.

Heart Conditions Can Result From Mineral Deposits

Calcium deposits in the heart chambers and valves become so cemented into place with mineral deposits that heart surgery becomes necessary. Calcium deposits in the inner ear cause deafness. Skilled surgeons can now remove these deposits and in many cases restore hearing again. Should these deposits occur on the nerve, then total deafness can result.

Since water is a carrier of inorganic minerals, harmful to the body functions, we now wisely remove virtually all trace minerals, chemicals and foreign particles through the process of distillation. Then, and only then, is water odorless, colorless, and tasteless and fit for consumption.

I realize that there have been a few university study results that seem to deny these observations. And it may be possible—in rare circumstances—that water can contain organically complexed minerals that are assimilable. But these are rare situations, especially on our now polluted planet. I am also aware of a study that tends to correlate heart health with hard water areas. Full scrutiny of these data, however, poses more questions than it answers. Not answered at all is how people are to get unpolluted water in the first place.

Pure Water Is Your Lifeline

We have now learned that water is a carrier of inorganic minerals and *toxic chemicals of organic synthesis*. But through the process of distilling, the water is separated from virtually everything it carries, and is now pure water. As it enters the body, it again picks up mineral deposits accumulated in the joints, artery walls, or wherever such deposits occur, and begins to carry them out. Gall stones and kidney stones get smaller and smaller until they can safely pass through their ducts. Little by little arthritic pains become less as joints become more supple and movable. Arteries gradually become more elastic as blood pressures tend to become more normal. Gradually the outlook on life becomes a little more youthful—ambitions begin to return, while the squeaking rocking chair will give way to the web covered golf clubs.

The startling fact to remember is that water attracts chiefly *inorganic* minerals. Organic minerals stay in the tissues, where they belong. This is a marvelous feat of Divine Reasoning. We were not born to die—we were born to live! This fact is easily proved. X-rays of the arteries will never show unless there are calcium deposits along the artery walls. Doesn't this prove that we have mineral deposits even along swift moving blood channels?

So hard water carries inert minerals into the body, and distilled water carries them out—it's just as simple as that.

There is then only one way you can purify your body and help to eliminate your chronic aging diseases and that is through the miracle of distilled water. A sludge-filled motor cannot operate smoothly unless the sludge is removed. Neither can a body be supple and ageless unless the joints, arteries, cells and nerve tracts are free of mineral deposits!

The Appalling Truth

The average person drinks about a gallon of water per day. Adding up the cups of coffee, tea, soft drinks, food and water, this gallon isn't too much. Many men drink much more. At a gallon a day, the average person drinks up to 450 glasses of solids during a life span. These same solids are found in your humidifier, and in grandmother's teakettle. Think of it—450 glasses of mineral solids in your system during a lifetime.

What does the body do with these foreign materials? Nature does the very same thing you are now thinking. It tries to eliminate them. Do you recall the water spots on your windows from your water sprinkler? Ever try to remove them? It takes lots of vinegar and lots of rubbing to dissolve the deposits.

This same reaction takes place in our bodies. It begins by putting a thin film along our intestinal walls. As soon as one film is laid, it attracts another film much easier, until it begins to build up and one of the first results is constipation, a plague to millions of people. Other deposits occur where blood flows the slowest, such as in the joints as arthritis and gout, along the arteries as hardening, in the veins as varicose, in the lungs as emphysema. It often coats the crystalline lens of the eyes with a fine film possibly resulting in cataracts. Glaucoma, the dreaded eye disease, can be another result of hard water. The tiny vessels film up with mineral deposits, which result in a build-up of pressure in the eye. There has never been a known cause for glaucoma. It might be the mineral deposits.

In all my years of experience as a practicing optometrist—all my glaucoma patients drank water with extreme hardness. Distilled water as a possible preventative is still the best long-range program to follow.

Heart Attacks Unknown

Bantu African tribesmen are known to have the cleanest arteries, as well as the most elastic ones. They dig no wells, but catch rain water. Although this water is contaminated with bacteria, it is still almost free of minerals. The air and water pollutions, such as we have here, have not as yet affected them as much. This is a tremendous testimony for mineral-free water. Even the Europeans have far less cholesterol in their arteries than we do. They of course drink more wine than we do. Wine is comprised of distilled water. All water that passes through the roots of plants, converts the inorganic minerals to organic minerals, and distills the water at the same time. When juices contain minerals, they contain them as complexed organic molecules.

The True Function Of Cholesterol

Cholesterol is very important to our system. We get it in certain foods. The largest part originates in saturated fats, and this type can be harmful. The liver also manufactures its own cholesterol. Cholesterol is an oil or lubricant. It keeps the blood oily so the blood can flow easily through the arteries and veins. If it were not for cholesterol, the friction along our artery walls would soon wear them through. Nature is not so stupid. Everything is planned to work. But as inorganic mineral deposits adhere to the artery walls, cholesterol collects, narrowing the diameter of the arteries, risking possible occlusion. The Bantu Africans did not have this problem. The best time to avoid mineral deposits is at an early age. This reasoning is logical.

My Personal Experience

About forty years ago I felt the impact of the effects of drinking hard deep well water. I never gave it much thought until recently, perhaps, because I was not concerned about longevity then.

This particular farmer had a most distressing health problem. I doubt that he weighed a hundred pounds. At intervals he would be unable to eat. If he did, he would belch, vomit and double up with pain. I thought each day would be his last. Finally his doctor advised him to leave his farm. Upon arriving in California, a kind neighbor suggested he drink distilled water with several small glasses of wine in between. A month later symptoms vanished and he could eat heartily. Today at the age of 88 he is still hale and hearty, an example of good health. It was the iron and magnesium in his well water that could have spelled an early death for him.

I Followed The Same Path

I still remember how I used to enjoy drinking ice cold water from his deep well, not realizing I was consuming three glasses of solids per year. That I too was a candidate for hardening of the arteries, arthritis, cataracts, gall stones, diabetes, obesity and other aging diseases never crossed my mind.

I then recalled another experience during my teen-age years. I was born and reared in a small South Dakota town. In those days we had a town well to supply our drinking water. It was good cold hard water. (It did a wonderful job in corroding my mother's teakettle.) The city electric plant was located near the well, but I noticed the plant manager drank distilled water. Being curious, I asked him why he drank that filthy water with such good tasting well water nearby.

I can still feel the indignation that suddenly swelled up within him. He then softened his words by saying, "Some day when you grow up, you will find out." Somehow this always bothered me. I knew he didn't want to take the time to explain it to a mere kid. All the while I knew he had his reasons. It took me all these years to find out.

The truth of drinking distilled water is not new. It is just that the public has never caught on. Today, many progressive doctors prescribe distilled water for their patients. All the kidney machines operate on the purest distilled water. It is now time to shake our sickly population out of their lethargy and languor. Thousands of American cities have potential health hazards in their drinking water and millions of farm wells should be condemned.

The Dark Clouds Are Gathering

Not only are we concerned about the minerals in our water, but we must also add the deadly nitrates and anthracenes which are seeping into our lakes, rivers and wells. The sewer wastes, and factory pollutions, detergents included, travel upstream underground at the rate of a half mile every six months.

Our President Shocked

Our presidents realize that through pollution we could destroy ourselves within the next few decades. Only through the backing of our government and by spending billions of dollars can we even begin to cope with this insurmountable problem.

What our presidents don't realize is that the first step to take is to purify our drinking water. According to government statistics, *10 parts per million of nitrates* in water can kill a newborn baby. This is serious! To put filtering plants in our cities would take millions of dollars. The quickest solution to our drinking water problem would be to distill and bottle water for delivery to our homes.

My Personal Problem Solved

I investigated this product and found distilled water was selling for about fifty cents per gallon a few years ago. It is higher now. I then began a search for a home still. There were many on the market for our laboratories, but the plumbing and the upkeep would be expensive, as well as a cleaning problem. Also most stills were not made for drinking purposes.

As luck would have it, I found an all-electric automatic still which just answered my purpose. The size of the still depends on the number of gallons needed. They are automatic and attractive in appearance, operate on pennies a day, and usually pay for themselves in nine months. It solved my immediate drinking problem for me and my family. I can now enjoy the purest water at its best!

We Are Our Own Worst Enemies

Now I can fight pollution. I can enter government programs raising my voice at town meetings against pollution in the soil, in the air and in the water. Today, the newborn baby doesn't have a chance if there are nitrates in its first bottle of water. Blue babies are the results of nitrates. How many babies are crying today when the hidden cause lies in the water, food, and the air they consume. Any mother would sacrifice her life for that of her child—yet unknowingly, through pollution—she becomes a hazard to her own child.

We Must Declare War On Pollution

The fight to end pollution calls for the combined efforts of every man, woman and child. The farmer must watch the insecticides and the toxic chemicals in his soils. The manufacturer must bottle up his waste products and eliminate deadly monoxide gas. The air must be made free and clean of all harmful particles. Pollutants are serious today—very serious.

Although I have solved my own problems so far as polluted and mineral saturated water is concerned, what about the millions of people in our cities who are unknowingly drinking water saturated with toxic chemicals and inorganic minerals? What about our rural people, who consider deep wells the epitome of good drinking water, not realizing the surgeries and aging diseases that may lie ahead?

Water Is Never The Same

True, our officials inspect and test our city water periodically, but water is never the same at all times, neither are wells in the same locations. Water can vary every day. Sometimes chemicals in water can become excessive; at other times negligible; but at all times the minerals will be there. We are looking for a mineral-free and chemical-free water. This is the only kind of water our bodies demand. Only through the process of distillation can we achieve this purity.

Water: A Carrier Of Disease

I recall a Dr. T. R. Van Dellen, who made the statement that even foul water carries the germs of hepatitis. In one ten-year time frame while I practiced in my profession, there were 1,700 cases of hepatitis due to contaminated water. Life is too precious to victimize so many innocent people.

House Beautiful magazine once made this statement: "You may still be able to drink water from your tap, but chances are it's not as pure as it could—and should—be. The pollutants in much of our water include chemicals from industry, such as lethal chlorinates, obnoxious phenols and tannins, plus commercial chemical fertilizers, weed killers, poison pesticides washed into streams by runoff or put into the water by injudicious overspray.

"Radioactive materials may one day be our most serious pollutant. Should atmospheric testing be resumed, this would be a menace. Also present is radioactive cesium-137 and even benzene-5, a known carcinogen.

"What can we do about it? We can improve our own home water supply. Water from our municipal systems has been treated to kill contagion, but still may not be appetizing. It comes from a sewage-polluted river or lake, it may be cloudy. It may be so hard that soap won't lather, but instead forms curds that puts a dirty ring around our bathtubs and turns our white clothes gray. If our water comes from a well, it may have all these faults and more. If you have a well, you may be floating on a sea of sewage.

"Now the day is coming [for many it is already here] when we, too, must face up to the question of whether we want to drink our water as it runs from the tap. Stopgap measures could correct some of its evils. But only *some*. The real answer lies in our rivers, our lakes and in our soil—and in what we do to protect them against befoulment. If we do nothing, a scourge of major proportion may be our legacy." So wrote David X. Manners.

This will give you some idea of how serious our water problem is! To clean it will take years. The answer lies in what we do *now* for our *own* protection. La Vista, Nebraska, had a severe water problem midway during my practicing career. The tap water collected in jars was rust colored and murky. Mrs. Carol Petersen, a local, said: "We had a lot of other neighbors who wanted to come here this morning, but they couldn't because they had to stay home with their children who had diarrhea from drinking the water."

Since water is a carrier of minerals and chemicals, let me list some of them found in our water:

Chlorine	Copper
Sodium	Zinc
Magnesium	Lead
Sulphur	Selenium
Calcium	Cesium
Potassium	Uranium
Bromine	Molybdenum
Carbon	Thorium
Strontium	Cerium
Boron	Silver
Silicon	Vanadium
Fluorine	Lanthanum
Nitrogen	Yttrium
Aluminum	Nickel
Rubidium	Scandium
Lithium	Mercury
Phosphorus	Gold
Barium	Radium
Iodine	Cadmium
Arsenic	Chrome
Iron	Cobalt
Manganese	Tin

Our bodies need many of the above-mentioned elements, but only in their organic state. They must first pass through the roots of our plants before our bodies can assimilate them. Herein lies the big error. Nature has a tremendous distilling plant. As the vapors rise, all impurities are left behind. Only as the clouds recondense the water does it again pick up the elements left behind. Now it becomes harmful to us, but beneficial to plants. There is a big difference between powdered iron filings and the iron found in plants. It takes no chemist to tell the difference.

Why Should We Grow Older?

The body is designed to feel younger—at any age. To feel one day older has always plagued the thinking of humanity and added worries to everyday living. New wrinkles add up to deeper lines—gray hair begins to crop out—while deaths occur at younger ages. All this need never be. Youth should remain youth—age is just a postponement—it need never come. The government is planning to educate 10,000 additional doctors to take care of the diseases for which we already have the natural cure. Why not attack the water problem, and prevent the host of diseases, without the need of additional doctors—who will die of the same diseases they attempt to cure! There is no need to have a heart so corroded with calcium deposits that surgery becomes necessary.

What A Price To Pay

One morning two patients came to my office. One had a ruptured blood vessel due to hardening of the arteries. He also had polar cataracts. The other patient had a severe nosebleed which resulted in the loss of a pint of blood. When he arrived at the hospital the doctor diagnosed his condition as arteriosclerosis, or hardening of the arteries. Had his nosebleed not relieved his pressure, it could have resulted in a stroke. Both patients lived on a farm, drank water excessively from hard water wells. Is it worth the price?

What Lies In Store For Us?

Could all these aging, crippling diseases have been avoided? What do we have to look forward to? Crippled joints and massive strokes? Do we want medical costs to wipe out our life savings? What lies in store for most of us?

The Rewards of Prevention

Now is the time to shout the truth from our mountain tops. It's time for the press, radio and television to splash this possible secret to longevity on their front pages, through the air and on our television screens. Let every man, woman and child become alerted to one of the greatest discoveries of mankind. Let our medical schools and practitioners turn their attention to prevention. Vital organs are not meant to be removed, but to function! Prevention, not drugs, is the answer to our aging problems.

Water Is Deceiving

What is in a sparkling glass of water may look pure, taste pure, smell pure, but you will still have to drink it at your own risk.

Here are pertinent suggestions:

1. The water must flow from the purest possible source. Adequate safeguards must prevent pollution of this source.
2. Frequent surveys must be made to detect potential health hazards.
3. An enforceable system of rules and regulations must prevent development of health hazards.
4. Proper safeguards must protect water quality from source to house.
5. Only qualified personnel may operate the water supply system.
6. The water source must have enough reserve to meet peak demands.
7. Bacterial monitoring must be continuous.
8. Only specified, scientific tests may determine water quality.
9. Water with bacterial levels above precisely stated limits must be considered unsafe for public consumption.

Consider, for example, Cleveland, which dropped thousands of pounds of nitrates and phosphates into Lake Erie, part of its metabolism. These wastes originated from homes, hotels, laundromats and offices. These wastes included, among other things, phenol, iron, zinc, sulfuric acids, ammonia, and hydrofluoric acid (which incidentally eats through glass). All this created complicated drinking water problems for Cleveland. And for most industrial cities.

It's Time To Become Alarmed

We do not know from day to day just what pollutions may exist in our own water. There is, however, one way to be sure, and that is to distill your own water. You are then sure that you will get the purest water possible.

A Twin Cities Problem

I remember reading about a Coon Rapids, Minnesota, housewife who telephoned the state health department to report that water from her home well had a peculiar "foam" on it. That was the first sign of a problem that later became one of concern to several million people in the Twin Cities area.

The "foam" came from surfactants, the active ingredient in detergents. Surfactants do not occur in nature. Thus, when they are found in an underwater supply, their origin can be traced directly to sewage discharge.

Nitrates become hazardous to formula-fed infants. Boiling water does not make the water safer, it only increases the concentration of the nitrates.

Nitrates—A Nebraska Problem

Nitrates fed to our Nebraska soils—to increase our food supply along the Platte Valley—have become a health hazard to the longest-lived people in the nation. Since the water table is high, nitrates filter into the water supply for drinking purposes. Since 10 parts per million can be fatal to a newborn child, it is time to become concerned. Our luscious valley, with its tall, waving corn, is rapidly becoming a death trap.

Today, mankind has destroyed our soils, our water and our air. The lure for wealth has over-shadowed our gift of life. Through our soils we have opened the doorway to hosts of new insects and plant diseases. Through our contaminated waters we are killing our wildlife and our babies in their cribs. The air is unfit to breathe. Life today is not living. It is merely existing. Where will it all end? Is it a problem for one of us, or all of us?

"If our soils are sick, our plants are sick. If our plants are sick, our animals are sick and if our animals are sick, we humans become sick," says Paul Brinkman.

Behind The Curtain

Americans during any year eat millions of pounds of beef from cattle that had "cancer eye," or similar tumorous disorders. Thousands of cattle carcasses checked by federal inspectors are held in meat plants until tumorous parts are whittled out. The remainder of the meat is put on the market.

Millions of cattle carcasses are detained briefly in meat plants while parts are cut out because of other diseases and injuries ranging from adhesions to tuberculosis. The rest is just "fine" for consumption.

In the meantime, federal regulations continue to permit meat processors to carve out infected or damaged portions of animals, condemn the cancerous parts of cattle, and sell the rest of the meat on the market.

Since the meat processors do not want to lose their profits, they would much rather feed this cancer-ridden meat to the unsuspecting public as "U.S. Choice."

Eco-Agriculture

I am happy to report that a new form of clean agriculture is gaining a toehold in the United States. It is being called eco-agriculture by the publisher of *Acres U.S.A.*—a journal layman and farmer alike ought to read—and it relies on scientific principles as well as the practical experience of organic growers.

Science has proved that plants with balanced hormone and enzyme systems provide their own protection against insect and fungal crop destroyers. The seat of this balance is soil fertility, starting with a pH managed by keeping calcium, magnesium, potassium, and sodium in proper equilibrium. The trace minerals also figure since they are keys to proper functioning of plant enzymes.

The full package known as eco-agriculture can and does produce more abundant bins and bushels, and does it while improving nutrient quality in food for animal and human consumption. What a boon to agriculture and to the health of our nation.

The people of Hunza do not use toxic chemicals. They grow from two to three crops per year. There are no insects or diseases. Why not channel our research and educational dollars where they will do some good?

The Wheel Of Health

We live by the law of transformation. Transformation from the soil to the plant—from the plant to the animal or human, then back to the soil again, forming an everlasting cycle of health. The dust of the earth then becomes the cells in our bodies. We live by the law of interdependence. No soul or body is an island by itself. It is a part of something else—always.

Therefore, we must first consider the soil, which is the mother of us all. We must become concerned about the air above us, from which we cling to the "breath of life." The water around us must be pure and free of all inorganic minerals and chemicals. Every cell in our bodies must be bathed in the purest water. Life is purity. We cannot live by "bread alone." Air, water and sunshine are responsible for life and growth. Each of these gives something to the other.

Paul Brinkman, a former president of Queen Ann School of Cosmetology, so ably expressed himself in these words: "Water cannot be destroyed, but it can be made sick! Man with his additives has defiled the laws of God. You can always expect something unnatural whenever you violate something natural."

Mr. Brinkman became interested in distilled water especially since he was concerned about his diabetic condition. Imagine his surprise when he checked out sugar free. It likewise surprised his doctor. According to any edition of *Merck's Manual*, the exact cause of diabetes is not known. The cause lies in the inadequate production of insulin by the beta cells of the islets of Langerhans. The cells are alive; it is just that they do not function. My inspiring teacher, Dr. Landone, stated that the mineral deposits enclose the cells with a mineral film so they are unable to function. I am inclined to believe his conclusions because there has to be a cause. Whether distilled water will cure diabetes is not for me to say. It is not within my field to diagnose; however, with the knowledge derived from my study of distilled water, I feel quite emphatic about its merits.

There is much to learn about water. I have merely touched on a few pertinent facts of water in its relations to vibrant health and longevity. The facts I gathered have proved themselves to me. When men can live to the age of 120, I become alert! There has to be some cause. I am convinced that the inorganic minerals and chemicals in water become a hazard to every man, woman and child. The body must be free of sludge to operate efficiently. The eliminating organs cannot expel all the deposits, so nature does the next best thing and tucks them wherever she can. Since water is the greatest solvent known, it becomes the best agent to carry out that which is brought in.

The test is very simple. Place a mirror or glass under a dripping faucet. Let it dry, and observe the water spots. Then place this glass in distilled water. The distilled water will dissolve the water spots. This is what the whole story is about—this is the cause of all our aging diseases. Dr. Landone cured his heart condition. Captain Diamond cured his arthritis and lived to the splendid age of 120 years. Could distilled water be the answer? If it is, how different living in this world could be. Since we live but once, therefore, should we not live wisely? The choice is clear—it is now up to you.

Disease does not exist without cause. It begins where cause begins and persists where cause persists. Chronic disease means chronic provocation. Cure cannot take place until the cause of disease is removed. Symptoms may subside or be palliated and suppressed; a crisis may pass, but disease persists so long as cause is ignored or unrecognized.

Let us now answer the many questions that may arise:

QUESTIONS AND ANSWERS

During my years in optometric practice, many of my patients have asked me the following questions. Naturally, I am not a medical practitioner, neither is it in our code of ethics to diagnose. As a profession, we draw a strict line between observing and referring. Therefore, I shall quote medical authorities, along with some of my observations. I only ask that you accept that which you believe, and disregard that which you disbelieve. The questions will all relate to "hard water" and its relation to our aging diseases.

Q. *In regard to your visit to Hunza, in your opinion, what was their main secret of longevity?*

A. There is no secret to longevity. However, I did observe the following. They ate most of their fruits and vegetables raw and raised them on organic soil. Fruits and vegetables are about 90% natural water of exceptionally high purity. Along with that, they drank glacier water which is very low in inorganic minerals. Wine was their main beverage, which again is comprised of natural water of very high purity. So their percentage of distilled water quality was 90% greater than ours. I consider this an important secret which I nearly missed.

Q. *Do you advocate raw fruits and raw vegetables?*

A. Indeed I do. A live body should have a live food! Plant a carrot and it will grow. Cook a carrot, then plant it. It will not grow, because its life's principle has been destroyed. Place cancer cells in a cooked media and they will proliferate. Now place the cancer cells in a raw matter and they will disappear! This astounding fact is known to the Cancer Society, but the public was never informed. By all means, get a good vegetable juicer. Through trial and effort I located a tremendous juicer—it answered my needs for a lifetime!

Q. *Will distilled water cure arthritis?*

A. According to medical science, there is no sure cure for arthritis. Arthritis is partially caused by mineral deposits in the joints. X-rays will show the mineral deposits very clearly. These deposits crystalize to irritate the joints as they move, causing severe pain and swelling. Since the average person drinks up to 450 glasses of mineral solids found in hard water during a life span, it is logical to assume that these inorganic minerals were slowly deposited in the joints through the years, causing the arthritis. Distilled water, in turn, dissolves these mineral deposits. Several years ago, I, too, was arthritic. At the age of 70, I could still kick as high as my head. It feels wonderful to be agile, and painless. Paavo O. Airola, N.D., in one of his early books, states: "There is a cure for arthritis." He reports the case in a Swedish clinic: "Albin Vistrand, a Swedish farmer, was so crippled by arthritis he could hardly move his arms and legs. Twelve years of treatment with prosthetic devices, drugs and X-rays had availed nothing. Yet, after only one month at the Brandals Clinic, his pain vanished, his limbs regained mobility and he went home to work on his farm—completely cured."

Q. *What was their method of cure?*

A. Live fruit and vegetable juices, fasts and warm baths with gentle massage. Again, your live juices are distilled water. The principal point to remember is that hard water carries inorganic minerals into our bodies, whereas distilled water carries them out. You must reverse this process. It's that simple.

Q. *What is your opinion of Dr. Jarvis and his honey and vinegar treatment for arthritis?*

A. It's wonderful. Only he should have suggested distilled water instead of hard water. Apple cider vinegar is a tremendous solvent. Along with distilled water, you have an unbeatable team. Add a tablespoon or two of raw apple cider vinegar [procured at specialty food stores] to a gallon of distilled water to give it flavor and you have a drink fit for the gods. Vinegar removes water spots from the windows, which proves what an excellent solvent it is. As to honey, it is predigested natural sweet. It will lessen your craving for synthetic sweets. Incidentally, raw honey is most excellent for allergies. Dr. W. Peterson, Ada, Oklahoma, said that raw honey is an effective cure for 90% of all allergies, providing this raw honey is obtained within a ten-mile radius. The above fact is most important! Dr. Peterson had 22,000 patients across the nation who were using raw honey. Only raw honey is of any value. Raw honey is available through local beekeepers, health food stores, or at some supermarkets. If it sugars—it's raw honey. Now, you can say "goodbye" to your allergies!

Q. *I have hardening of the arteries. Is there a cure for it?*

A. Medical science would give a qualified "no." They classify it as an "aging" disease. Arteries are invisible to X-rays unless there are mineral deposits along the artery walls. They then show up very clearly. This proves one point, that mineral deposits do occur along the artery walls. That is the reason they "harden." Also, this is another reason why they lose their elasticity and rupture after a sudden exertion. Could it be if we switched to distilled water at an early age, this dreaded condition would never exist? Captain Diamond thought so, and lived to the biblical age of 120 years. A close friend of mine hoped to have a coronary artery transplant. When the surgeon examined his coronary artery, he found it [in his words] "hard as cement." There was no way to attach an elastic artery to a "cemented" artery. Could distilled water during his younger years have avoided this? Dr. Landone's experience says "yes."

Q. *Does hard water affect everyone alike?*

A. Definitely not! Of the 106 different chemicals and minerals found in water, each person may become acceptable to one or more, or a combination of them. Mineral deposits settled in different parts of the body, causing different symptoms. If mineral deposits film the intestinal walls, it could result in constipation. If they clog the screening system of the kidneys, it could result in a kidney transplant or kidney stones. If they corrode the heart chambers or valves, a transplant could be necessary. Water rushing swiftly through a pipe can still corrode the pipe. Speed makes no difference. A glass of water may look clear and sparkling, but spill it on a mirror, let it dry, and film remains!

Q. *How do you determine hardening of the arteries?*

A. How does the physician determine when a person has hardening of the arteries? The retinal vessels are seen by looking into the eye with an opthalmoscope. Arteriosclerotic vessels are corkscrew in appearance. Whenever a hardened artery crosses a soft vein, the latter appears to be indented or nicked. A diagnosis of arteriosclerosis can be made in this situation. X-rays of the chest, abdomen, or extremities offer reliable evidence. Thickened and tortuous vessels are demonstrated on the film when there is calcium in the walls. The physician assumes that arteriosclerosis exists in older people and most of us probably have a patch here or there. Traces have been found in postmortem examinations on infants and teenagers. Calcium deposits do exist within our artery walls. "Most heart attacks stem from blockage of an arteriosclerotic coronary artery. This is true especially of persons suffering an attack of coronary thrombosis. These symptoms are direct evidence that the condition exists. We do know that the coronary arteries are most susceptible to the hardening process. As a result, they may be involved long before those of the brain, eyes, kidneys or limbs."

Q. *Does lecithin have any merits?*

A. Lecithin is a fatty acid found in animal tissues, especially nerve tissue, semen, yolk of egg, and in small amounts in bile and blood. Lecithin is said to have been given in rickets, dyspepsia, neurasthenia, diabetes, anemia and tuberculosis, according to a standard *Medical Dictionary*. Lecithin also acts as a solvent for gall stones. It would then be wise to combine lecithin with distilled water for greater effect. Lecithin can be purchased in capsule form from most natural food stores. This is a good preventative before gall stones form. It is then important that you remain under doctor's care to avoid any blockage. Your doctor is a very important consultant for your health problems.

Q. *Surely tested city water ought to be good to drink!*

A. Grand Island, Nebraska had a severe water problem when I was an area practitioner. When a local meat packing plant was closed down, due to staining of meat by iron and manganese, the seriousness of the problem was brought to the attention of the city. The cry then arose, "What will this water do to children and adults?" The cost of a 50 megawatt steam generating unit was about $17,572,937 at the time. What a price to pay to remove the minerals from their drinking water. How much better to have a home distilling unit now.

Q. *Is fluoridation healthful or harmful?*

A. Dr. Robert Mick of Laurel Springs, New Jersey, has offered $100,000 to anyone who could prove fluoridation is healthful! At first an ardent believer of fluoridation, he began experimenting with animals. As a result, he learned that bones, teeth, kidneys, livers and spleens had accumulated up to 500% more fluoride than controlled animals. Cripples were born to the third generation! His reward has never been challenged. Dr. E. H. Bronner, Escondido, California, a nephew of Albert Einstein, offered $100,000 to anyone who could prove that fluoridation was beneficial to humans. Fluoridation not only hardens teeth, it also hardens the arteries and brain, and the same compound is used as a rat poison. An estimated 88 million pounds of rat poison shipped to our water systems for human consumption are not colored blue or marked poison. Only fluoride which goes to exterminators must be colored blue, or the supplier goes to prison. Fluoridation of public drinking water is criminally intolerant, utterly unscientific, and chemical warfare. Thank heavens, distilling water will remove these deadly fluorides!

Q. *What are some of the fringe benefits of distilled water?*

A. Tea has more flavor. Coffee requires one-third less granules. Ice cubes are crystal clear, foods digest much better and vitamins assimilate easier and become more effective. The body derives more nutrients out of foods and the corpuscles carry greater loads of oxygen, so important to cell life. Foods keep much better and longer [On soft water cattle produce 20% more milk and cream on 20% less fodder and the milk tastes much sweeter, with less bacteria. This is based on Emmett Culligan's research.]

Q. *What does distilled water do for emphysema?*

A. According to *Merck's Manual*, it is very difficult to diagnose all forms of emphysema. In all cases corpuscles cannot come to the surface of the lungs to discharge carbon dioxide and to return with oxygen. Even children have patches of mineral deposits somewhere in their tissues. Dr. Brown Landone, stated that often mineral deposits surround cells and imprison them so they cannot function. X-rays show such deposits along our joints, and along artery walls. Perhaps such patches, accentuated by heavy smoking, could so seal the surface of the lungs that a condition of emphysema could exist. The condition could be much like spraying kerosene over a stagnant pond to keep the larvae from coming to the surface. The theory advanced by Dr. Landone is that inorganic minerals do form somewhere in our bodies. In Dr. Landone's case, the minerals deposited formed themselves in his heart valves. After drinking distilled water for several years, his heart was perfect. He lived to the age of 98. Whether this applies to emphysema is not for me to say. Many readers have written to me about its merits. Since Dr. Landone lived to the age of 98, I had better lean in his direction.

Q. *What effect does distilled water have on cancer?*

A. Again, according to Dr. Landone, "Hard water seals each cell with a film, so oxygen cannot reach the imprisoned cells. Nature then develops new cells which thrive on less oxygen. These cells are called cancer cells. Distilled water often frees the imprisoned cells and allows the oxygen to reach the cells." Does it not seem logical that a lifetime intake of distilled water could possibly have very beneficial effects?

Q. *What can distilled water do for obesity?*

A. Dr. Landone stated emphatically that anyone drinking distilled water exclusively would eventually return to normal weight. The reason: hard water film imprisons the cells so tissues become water-logged. Retention of fluid is the chief cause of obesity. Distilled water again breaks the cell barrier and the body weight returns to normal. Dr. Landone kept his weight perfect throughout his 98 years. Weight watchers will lose weight, regardless of present diet, if they will include from four to six glasses of distilled water a half hour before each meal. It works wonders. Why not try it!

Q. *How can I drink distilled water on my coffee break?*

A. I had this same problem, but solved it quite simply. I have distilled water in my office. Before I go out on my coffee break, I drink several glasses of distilled water. I then order a cup of coffee, and sip a little off the top. Coffee breaks are for conversation and a change in routine. On a trip, I carry several bottles of distilled water with me. If I go on longer trips, I pack my portable electric still, and have my own water from any tap. When I traveled in foreign countries under the sponsorship of Art Linkletter, I contracted dysentery so badly I was disabled for several months. Finally, Campbell's tomato soup cured my dysentery. My water still could have prevented this. What a boon for world travelers! Good health and good sense are two of life's greatest blessings. We often wonder why streams are so bitter, when we ourselves have poisoned the fountain.

Q. *What was Connie Mack's secret?*

A. Connie Mack, manager of the Philadelphia A's for nearly 30 years, reputedly would not allow his players to drink hard water under any circumstances. He would not accept any contract unless he could furnish his men with distilled water, and on some occasions, low-mineral mountain water. Distilled water kept his players free from constipation and in perfect health. Seldom did his pitchers have sore shoulders due to hard water mineral deposits in their joints. If your aspiring pitchers consider the big leagues as a career, let them remember Connie Mack's secret. Connie, too, kept his perfect health to the age of 90.

Q. *Do babies need distilled water?*

A. Indeed they do! Many doctors insist on it. Babies are born free from pollutants, and small amounts of pollutants to an unadapted baby can be serious. Their formulas should include distilled water. Prickly heat or rash are caused by hard water deposits left on diapers and clothing. Often small amounts of nitrates can be fatal to a newborn child.

Q. *What will distilled water do for my complexion and scalp?*

A. I recall a large house-to-house cosmetic company that went broke. Their amazing success was due to one simple product: perfumed distilled water. A cotton ball was saturated with this preparation to clean the pores. The patrons were so excited about this preparation, they asked what was in it. As soon as they discovered the truth, they made their own. Hard water will seal the pores, much as it spots windows. These spots are visible under black light. No wonder we have so many skin infections, pimples and other eruptions! Switch to distilled water, with non-detergent soaps, and many complexion problems will vanish.

Q. Can animals and wildlife tell the difference?

A. Place the nine different kinds of water before a goat and he will pick out the distilled water. You can't fool him. Birds given distilled water will return year after year. Thoroughbred horses have lost many a race due to changes in water. The late great race horse, Secretariat, drank only distilled water. A breeder of thoroughbreds from San Diego told some amazing stories about his own horses. *Acres U.S.A.* has numerous case reports on the value of pure water in animal production. Whenever possible, distillation is preferred. If more race horse owners knew this simple secret, they could have fattened their purses many times.

Q. Does hard water affect appearance, taste?

A. Here is a curious twist in the food department . . . certain vegetables like peas, beans and legumes are effective water softeners! They absorb a certain amount of the mineral hardness in the water, becoming wrinkled in appearance. In turn they lose as much as 50% of the valuable nutrients [B1]. Coffee and tea are also affected in that not enough true flavor comes through because of the hardness content in water, which explains largely the reason for the different taste from one place to another.

Q. Some writers advocate calcium found in hard water. They say it is good for your heart, arteries and bones.

A. The answer is very simple. If calcium is in the heart tissues, in the arteries and in the bones, it is good for you. However, if it is found on the *outside*, it is definitely not good for you. Hard water deposits its calcium and other minerals on the *outside*. The choice is clear!

Q. Are others concerned about drinking water?

A. When I started my practice the only water-borne virus diseases were hepatitis and poliomyelitis. Today there are over one hundred! Sources in the literature would probably fill a booklet this size. The diagnosis of one river, the Connecticut, once revealed how foul our "fresh" waters have become. A random sample of the Connecticut disclosed disease bacteria including typhoid, paratyphoid, cholera, salmonella, tuberculosis, polio, anthrax, tetanus, plus countless other viruses.

And that is not all. Repulsive parasitic life forms such as tapeworm, roundworm, hookworm, pinworm, and blood flukes were also present in abundance." John Dvorak [*Omaha World Herald*, August 29, 1970] made this statement: "Fish placed in the river near Omaha, Nebraska, had an unacceptable taste. Fish from between the mouth of the Platte and just above Kansas City tasted 'acceptable.' However, in the Kansas City area, river pollution perforated the skin of the catfish." Water samples taken from the Mississippi River below St. Louis were found to be so toxic that even when diluted ten times with clear water, fish placed in the mixture died in less than one minute.

Q. Do we need the inorganic minerals found in our water?

A. Absolutely not! We get our organic minerals from the foods we eat. Fruits, grains, nuts, vegetables and meats supply all the minerals we need. The longest-lived peoples, such as the Hunzakuts in the Himalayas, the Vilcabamba in Ecuador, and the Abkhazia in Georgia (south Russia), live to fantastic ages. They drink glacier water, which checks out as single distilled water. If you would like a clearer understanding, let me put the gist of my book in a single paragraph. Pour a cup of tap water in a pan and boil it. The vapors that rise are pure distilled water. It is fit to drink. The spots that remain are inorganic minerals and chemicals—unfit to drink. Rain water will not spot your car. Tap water will. We do not want to drink water that "spots." Only the purest water free of all minerals and chemicals should enter our bodies. At one time our entire planet was covered with vapors, which condensed into rivers, lakes and oceans. As these vapors condense, they change into the purest water we can drink. A simple rule to follow is this: *"If water 'spots,' it is unfit to drink. If it does not 'spot,' then drink it."*

Q. Does distilled water leach organic minerals out of our systems?

A. Of course not! "It is only inorganic minerals rejected by the cells and tissues of the body which, if not evacuated, can cause arterial obstructions, and even more damage. These are minerals which must be removed and which distilled water is able to collect." So said Dr. N. W. Walker, age 100. In the business of telling people how to live to be 100, he has track record. *Living Health* by Harvey and Marilyn Diamond confirm this finding.

Q. What do the experts say?

A. Experts say it will take 25 years and $70 billion to make a dent in our world pollutions. For our immediate relief, we need to go back to our organic gardens, raise our own beef, and distill our own water! If we wait—it may be too late. The fire of pollution keeps burning and destroying. To survive, we must act now! The prophet Isaiah graphically foretold of our day: "The earth is drooping, withering . . . and the sky wanes with the earth, for earth has been polluted by dwellers on its face . . . Therefore, a curse is crushing the earth, alighting on its guilty folk; mortals are dying off, till few are left." (Isaiah 24:4-6. Moffatt translation) . . . "even the beasts and birds and the very fish within the sea are perishing." (Hozea 4:3). I quote these from *Our Polluted Planet*, Ambassador College Press, Box 111, Pasadena, California 91101. Dr. Paul Bragg, who lived to 94, a life-long exponent of vibrant health, wrote a startling book, *The Shocking Truth About Water*, which should be studied by everyone. No hard water had ever touched his lips. He had the blood pressure of a 20-year-old youngster. Let me quote his first experience with hard water. *"When I was a small boy, my father took me to the P. T. Barnum Circus. The most fascinating freak to me was the lady who had turned to stone. Here was a woman on a bed, and they could actually drive spikes and nails into her body. She was so full of arthritis and acid crystals that she had no feeling left in her body. She lay helpless and rigid. She could move only her eyes. This lady suffered with complete ankylosis—that is, there was no joint in her entire body that could make a simple movement. All the nerve tissues in her body were paralyzed and dead. The man who explained these freaks said that this lady was born in Hot Springs, Arkansas. The lady who had turned to stone was a complete mystery to me as a child. But not today! The water in Hot Springs is some of the heaviest water in the United States. I have seen analyses of it and the concentrations of calcium carbonate, potassium carbonate and magnesium carbonate are very, very high. The poor lady in the side show was a victim of this inorganic water. Her vital organs were not strong enough to help force those inorganic minerals out of her body, so they deposited themselves in her joints. This was an unusual case, of course, but I have seen many, many cases of arthritics who were complete cripples, absolutely helpless. There are more than 20 million people living in the United States today who have arthritis to some degree."* Dr. Bragg was a shouting testimony on how to live in a 20-year-old body for 90 years or more. Was distilled water one of the factors? The choice seems clear! Why wait for crippling and aging disease? Why not start while young and avoid that which is sure to come? Youth is not measured in years—the future of everyone can be glorious, vibrant, and full of the adventure of achievement. Continued youth is within the reach of all hands. Health is like the air we breathe. We don't miss it until we are deprived of it. "Personal liberty is not a license to do whatever you wish, but freedom to do what you ought."

Q. Why didn't so many of the early practitioners write down what they had to say rather than pass on their observations to you verbally?

A. Thirty-five years ago a writing that contradicted official science was a one-way ticket to prison. Victor Irons went to prison for what he wrote about *Green Life*. Wilhelm Reich died in prison for his findings on energy outside the electromagnetic spectrum. The late Dr. Carey Reams spent his life and fortune defending his biological theory of ionization and spent time in prison just the same. Only recently has this judicial tyranny been lifted somewhat. We owe citations to many people, even though they didn't write.

What Medical Authorities Say

H. W. Holderby, M.D., in his amazing *Report On Water*, has this to say:

Now a new poison has been added to our drinking water and toothpaste. It threatens our health and lives. Millions of people are slowly poisoning themselves to death.

The medical and dental world has just been rocked to its very foundation by government sponsored investigations of toothpaste containing fluorides. For fluoride is derived from fluorine, a deadly chemical to humans, and while not dangerous in small amounts, the government investigations of 156 cancer deaths which occurred in the past three years indicates that fluorine accumulates in the human body and eventually causes cancer and/or other illnesses which are fatal. Toothpaste is the second largest source of fluorine, next to water which is Number One.

Fluorides attack almost everything. Their chemical action is such that all containers must be lined with rubber or plastic, for it eats through all metal material. What must it do to the tissues of the body? It is called "The Wild Cat" of the chemical world. It unites with more chemicals than anything else, forming new compounds. How can it help the teeth, which are made up of calcium, when according to the U.S. dispensatory (24th Edition, page 1456): "Fluorides are violent poisons to all living tissues, because of the precipitation of calcium."

Medical authorities now state it can cause cancer, Mongoloid births, kidney disorders, bone diseases, impotency and even madness.

A copy of Dr. Holderby's report is available from Natural Foods Associates, P.O. Box 210, Atlanta, Texas 75551.

What Consumers Say

A poison bottle has a skull and crossbones on it, but there was no skull and crossbones on the faucets in Madison when I came here to tell me that I was drinking and cooking with one of the most deadly poisons known to man. It was not until 16 years later that I learned about Madison's fluoridated water and could take remedial action. Sixteen long years of fluoridating my system, seven of them wracked by pain and despair because I did not know what was wrong and my doctor could not help me.

I've studied every book and article I could find on fluoridation. I've talked with those who know fluoridation. I've been to the health authorities—the men who should have helped me—and was told to move out of Madison if I thought that fluoridated water harmed me. Mothers would not feed their children the fluoride tablets which would strengthen their teeth against cavities and so the children would have to be doctored whether they wanted it or not.

The question most often asked is, "How do you know that fluoridated water harms you?" That poses a problem, for how do I know? If, every time I ate an egg it decided to return the way it came, I'd know that an egg was taboo in my diet. Or if I were to drink some lye water and burned myself from end to end, I'd know that my body says, "No, no! Mustn't touch!"

But how do I know that fluoridated water is a no-no? There's no warning burn, the stomach does not rebel (at first); there's no smell, no taste to warn of danger. So how do I know?

Only by bitter experience do I know, and that is a long, long story. I'll just hit on the high spots.

I was the Paddock tomboy. I could jump and run, climb trees, swing as high as anyone else. I spaded the garden, cleaned the hen house, loved the outdoors. I joined the WACS and marched along with those much younger than I. I went to Alaska and helped my brother-in-law build a garage. I was healthy.

Oh, yes, I had my troubles. My days when my stomach rebelled against certain foods (a dormant ulcer): the days my back ached from the slipped disc which had somehow come into my life. But, all in all, I was a healthy individual. And then I came to Madison! Gradually that health gave way to aches and pains, to allergies and sleepless nights, to intense nervousness.

When I came to Madison in 1949, I did not know that Madison's water was fluoridated. I did not know why, suddenly, my dormant ulcer should start kicking up, why there was constant pain in my right side, why cold chills ran up and down my left side when I lay in bed. Neither did my doctor know, for he'd never been taught to say, "Possibly you cannot tolerate fluoridated water. Let's quit it and see." Instead he said, "Getting old, Mona, getting old."

No, it was up to me to find out what had happened. It was up to me to determine why I was so sore that I could not turn over in bed without pulling myself over by grasping the edge of the mattress; why I could not sleep through a night without such pain that I must spend the last three hours in the reclining chair. It was up to me to find out why I had suddenly developed allergies; and to determine that after six months without fluoridated water I no longer needed the shots which had been given me on a regular basis for over seven years. And I found it out by myself—just by not using Madison's fluoridated water.

Washing dishes became an impossible task, for my lungs and ribs seemed ready to break out in rebellion. The doctor thought the sink was too low; he told me my lungs were okay, there was nothing wrong with my liver—there was nothing wrong with me.

Then why couldn't I write? My job depended on my writing notes on the margins of the students' papers, and I found myself writing about five times the size letters I'd always written. As I wrote my arm would suddenly twitch and the stroke would jump up the page about an inch. It was aggravating and humiliating. Yet what could I do? I was okay physically, my doctor said I was.

Then I found out about Madison's fluoridated water and I quit drinking it. The Natural Foods people in Madison helped me to obtain food and water that was not fluoridated, that was not poisoned by fertilizers, herbicides, and insecticides. I owe a mountain of thanks to them and my only way of repaying them is to help let the world know about the damage fluoridated water can do to the health—mental and physical—of some individuals.

You may say, "But possibly it was the herbicides and insecticides which harmed you." Yes, they harmed me, but I've found out by drinking just one glass of Madison's water (not changing my diet in any other way) that the old pains and miseries will return. And so now I know.

Now I know that if I am to live in Madison—and my work demands that I live here—I must not drink its fluoridated water. Now I know that I must carry my own food and water with me when I am to be away from my apartment for any appreciable length of time. I have a special drawer in my desk at work where I have three small jars of water on hand at all times. I cannot go down to the break room and have a cup of coffee with my co-workers; I must sit at my desk for my break. I cannot go to office parties, church gatherings, or to the restaurants in Madison for a "snack" with friends during a shopping tour. I cannot go to conventions in some cities in Wisconsin without carrying my food and water with me. And where am I to eat while my friends are banqueting?

Recently I rebelled, decided that one meal with friends would not hurt me. I was wrong. It took me three weeks to undo the damage. My ankles remained swollen for that period, and are still tender to the touch. The old pain in the intestinal tract returned, my heart did a thump-thump at the least exertion, and my eyes watered, my hands trembled as with the ague.

I recently found it necessary to go to another dentist who told me, on my second visit, that my teeth were mottled. Mottled teeth? Why, I asked? He did not know. Nor do I.

The small Wisconsin town where I spent the first 19 years of my life contains about two parts per million of fluorine in its water. Was that sufficient to cause mottling? I suppose so, for the U.S. Public Health Service has found mottled teeth in Kewanee, Illinois, and Marion, Ohio, which both have naturally fluoridated water.

Could my jawbones have been affected by the fluorine I got from the drinking water as a child? Could that be why, after coming to Madison and drinking its fluoridated water my jawbones ached when I ate carrots and chewed meat? Could that be why large fillings fell out of my teeth after I came to Madison, why the dentist could find no reason for the constant pain in one incisor?

Could be, for that pain left me after I quit drinking Madison's fluoridated water; but the damage is there. Important teeth are gone and store-bought ones replace them.

The damage is there, too, for I cannot use Madison's water for any purpose—not even for bathing and for washing dishes. That statement seems incredible, but it is true. I suppose it is the steam which affects me. Some say it is possible; others say it is impossible. All I know is that my lungs and ribs are no longer sore. I sleep through the night, I roll over in bed with practically no pain. I write between the lines with small, compact letters—and my pen does not jump the page. My doctor says, "As long as it makes you comfortable. That's all that is necessary. Who is to say why you are affected, and the vast majority of people are not? Who is to say that you imagine things, and be right in their assertion? You are getting better every day. You're free of pain. Let them talk."

But I do not let them talk any more. I talk back. I'm fighting it now, out in the open. I'm locking horns with those who say I'm a kook, a food faddist, a hypochondriac. I may not see the end of fluoridation in my time—but they will know that I'm not silent. The end of fluoridation would not help me, for the remnants of it will be in the pipes for many, many years, and my body is so sensitized to it that I could not drink the water even if they never put another drop of fluoride into the water in Madison. Sad, but true!

So keep up your good work. I'll fight in my small way. Others will join forces. And one day we will lick it.—*Correspondence from Ramona H. Paddock, Madison, Wisconsin to Dr. Allen E. Banik.*

• • •

It might be of interest to you to know that I can substantiate your thesis on distilled water, *The Choice Is Clear*, which I have just read.

On a St. Patrick's Day many years ago I was hospitalized with a massive coronary thrombosis. I was "a fat old man" with all the "qualifications" for becoming a typical cardiac cripple.

The emergency procedures of the medical doctors undoubtedly saved my life, but I had serious doubts as to the efficacy of their long-range cures which included heparin shots, etc. Instead, I put myself on a strict regimen of diet, vitamin therapy, distilled water, and exercise.

The results have exceeded my expectations. I feel better at 67 than I did at 47. My arteries have softened; my joints have limbered; my vision is sharper; my nerves are calmer; and my head is clearer. Your book and Dr. Bragg's have convinced me that the distilled water has been the most important facet of my rejuvenation program!

Good luck in your fight against pollution. Your book should be required reading in schools and colleges.—*Ben H. Martin, California.*

What Others Have Written

"Seven years ago I broke my arm. Later on arthritis set in, and I was unable to raise my arm, or sleep on my right side. After reading your book and applying your theories, my arthritis disappeared, and I can now sleep most comfortably."—*Mrs. P. B., Nebraska*

"I was completely exhausted doing even a part of my housework. Now I can work all day and enjoy it."—*Mrs. J. L., Nebraska*

"I just finished reading your book, *The Choice Is Clear*. You spelled it out very clearly. I consider it a most excellent presentation. I would like to obtain these books in quantity to help spread the gospel."—*Pastor C. N. K., California*

"Your book proves one of the most exciting breakthroughs in healthful living. It made a believer out of me. Please send me a hundred copies to give to my patients."—*Dr. W. G. H., Iowa*

"Through the years I lost four parakeets with infectious arthritis. Pronounced incurable. This disease again struck the fifth one. His little feet were all crippled, and he was just wasting away. After applying your suggestions, my parakeet is now well and happy. The deposits are dissolved from his feet, and he can nimbly hop on his perches. He is eating again, to a point of being plump."—*Dr. J. C. T., Atlanta, Georgia*

"My doctor said, 'You have an incurable disease.' A month or so ago, my ankles began to swell. After drinking distilled water, the swelling went down. I am now beginning to feel better all over. I never realized the marvelous benefits of distilled water." —*V. E. C., Minnesota*

Soft Water And Hard Arteries

Some years ago, *Lancet* magazine stated there may be a correlation between soft water and hardening of the arteries and that "hard water" saturated with calcium and magnesium added elasticity to the arteries.

Some of our unwary writers, advocating the inorganic way of life, somehow twisted their thinking to the "hard water" theory. It is my firm belief that once you believe in the organic way of life, you better stick to it all the way.

Dr. Henry Schroeder, of Dartmouth Medical School, an opponent of the "hard water" theory, has long been intrigued by the physiologic and pathologic effects of trace elements. He lays the blame not on soft water—nor its lack of calcium and magnesium, but to its *high cadmium* content derived from the plumbing in the soft water supply. Cadmium competes with zinc to coenzyme processes, replacing zinc in metabolic activities related to fat utilization. He feels that *cadmium* is the key factor in the pathogenesis of both arteriosclerosis and hypertension. His excellent *The Trace Elements and Man* is now in print.

So it is not the soft water that is to blame, but in the chemicals water picks up through our water mains. Distilled water, or any water, must not travel through copper, iron or lead pipes. For a simple experiment, pour a little water in a glass jar and let it evaporate. Do likewise with distilled water, and note the results.

It is now well recognized that gallstones are formed from the constituents of bile and contain cholesterol, bile pigments, and inorganic calcium salts. If calcium contributes to gallstones and kidney stones, why would it not cause hardening of the arteries? The choice seems to be very clear!

This book was written to help you understand the relation of water to your health and longevity, and subsequently the health of our nation. As you plan your future and that of your children, don't overlook the underlying cause of all aging diseases. Learn the facts; become aware of the truth and build a strong, healthy America.

We do not live alone, nor do we work alone. Only by the combined efforts toward a common cause can we move the mountains before us. With this thought in mind, I humbly submit my small contribution to you. May it grow in your heart as it did in mine.